D. O. Nowikowa
A. F. Dulin

RESEARCH ON DORMANCY AND SEED GERMINATION AS AN EDUCATIONAL RESOURCE

AF293863

D. O. Nowikowa
A. F. Dulin

RESEARCH ON DORMANCY AND SEED GERMINATION AS AN EDUCATIONAL RESOURCE

FOR LEARNING AND RESEARCH ACTIVITIES OF STUDENTS

ScienciaScripts

Imprint
Any brand names and product names mentioned in this book are subject to trademark, brand or patent protection and are trademarks or registered trademarks of their respective holders. The use of brand names, product names, common names, trade names, product descriptions etc. even without a particular marking in this work is in no way to be construed to mean that such names may be regarded as unrestricted in respect of trademark and brand protection legislation and could thus be used by anyone.

Cover image: www.ingimage.com

This book is a translation from the original published under ISBN 978-620-3-04194-1.

Publisher:
Sciencia Scripts
is a trademark of
International Book Market Service Ltd., member of OmniScriptum Publishing Group
17 Meldrum Street, Beau Bassin 71504, Mauritius
Printed at: see last page
ISBN: 978-620-3-33825-6

Contents

INTRODUCTION

The Russian Ministry of Education sees one of the ways to improve the effectiveness of the educational process in a modern Russian school as the development of extracurricular education through elective and elective courses. The elective courses, in addition to teaching and research activities, may include elements of scientific and research work and, in this regard, it is important for the teacher to choose a topic, a direction within which students could work. One of such directions is the least studied in the physiology of plant growth and development problems of dormancy and seed germination.

The hormonal system of plants is known to regulate such important processes as seed formation, their transition to dormancy, and seed emergence from dormancy (Obrucheva, 2012). There is no doubt that gibberellins, auxins and cytokinins play an important role in some stages of seed formation (Nefedyeva, 2013). It is also established that these hormones and abscisic acid regulate the process of seed germination (Kefeli, 1989). According to M. G. Nikolaeva, indolylacetic acid is the hormone whose content determines the transition of seeds to the dormant state and their exit from it. Meanwhile, the role of such relatively recently discovered phytohormones as brassinosteroids, jasminic acid and salicylic acid in the processes of seed formation, dormancy and germination is not adequately studied (Dulin, 2010).

Scientific hypothesis: we hypothesise that salicylic acid and brassinosteroids take part in the processes of seed formation, seed dormancy, seed emergence and germination, the combined use of salicylic acid and epibrassinolide may contribute to an additive or synergistic effect.

The aim of this work is to study salicylic acid, epibrassinolide, indolylacetic acid and gibberellin when used separately and together on the processes of seed formation, dormancy and germination, the use of literature material and experimental data obtained in educational and research activities of students.

Objectives:

1)   Review the literature on the role of hormones in seed formation, dormancy and germination.

2)   Carry out an experiment on the effect of phytohormones on the quality and morphometric parameters of seeds formed from flowers of the mother plant treated with growth regulators.

3) Investigate the effect of exogenous hormones on the germination of isolated embryos.

4) Develop an elective course for grades 10-11 based on literature and experimental data.

Subject:

1. Seeds   and   isolated   germs   of   Manchurian hay,   cherry tree, Sargent's wort, Manchurian apricot.

2. The learning and research activities of the students.

Subject of the study:

1. Seed formation, dormancy and germination processes under the influence of exogenous phytohormones.

2. An elective course in biology for grades 10 - 11 on "The combined effect of phytohormones on the growth and development of seeds of Far Eastern plants".

Relevance:

1. In the literature there are single works on the influence of salicylic acid and brassinosteroids on the processes of seed formation, dormancy and germination in a limited number of objects - seeds with shallow dormancy, so the study of seeds with deep physiological and morphophysiological types of dormancy is relevant.

2. The relevance of developing an elective course is related to the need to provide high school students with an opportunity to broaden and deepen biological knowledge, develop skills and experimental work skills, assess their readiness to study a given specialty (biology) through the experience of studying specialised disciplines within their chosen field.

Scientific novelty:

1. The effect of exogenous hormones salicylic acid and epibrassinolide on seed dormancy formation of Manchurian ash, felted cherry, Manchurian apricot and Sargent's wort was studied for the first time. It has been found that treatment of flowers of mother plants of these species by growth regulators salicylic acid and epibrassinolide influences the dormancy depth of forming seeds and the growth of their seedlings. No additive and synergistic effects were observed when salicylic acid and epibrassinolide were used

together.

2. Experimental works for a school experiment on the topic of plant growth and development are proposed.

Scientific and practical relevance:

1. Plants (seedlings) of Manchurian apricot and felted cherry can be obtained from seeds without preliminary cold stratification, using the culture of isolated embryos using exogenous hormones salicylic acid and gibberellin.

2. An elective course in biology for grades 10-11 entitled "Plant Growth and Development" was developed on the basis of the research materials.

The statements made for the defence:

1. Treatment of the flowers of the mother plant with salicylic acid and epibrassinolide has a positive effect on dormancy and some morphometric indices of the formed seeds.

2. Exogenous salicylic acid and epibrassinolide in different concentrations promote the emergence of isolated Manchurian ash, felted cherry and Manchurian apricot germs from dormancy and enhance some growth processes.

3. Based on an analysis of the literature and experimental data, an elective course in biology for grade 10-11 entitled "Plant Growth and Development" has been developed.

4. The application of salicylic acid and gibberellic acid on isolated embryos of Manchurian apricot and felted cherry tree allows seedlings to be obtained under field conditions without cold stratification.

Work approbation. The results of this work were presented at the XVII International Scientific-Practical Conference "Problems of botany of South Siberia and Mongolia" (Barnaul, 2018) and at the XX International Scientific-Practical Conference

"Experimental and theoretical research in modern science" (Novosibirsk, 2018).

Publications. 2 articles were published on the topic of Master's thesis, 1 of them in the collection of scientific articles included in the RSCI database.

Structure and scope of work. Master's thesis consists of an introduction, 4 chapters, conclusion and list of references. Materials of the Master's thesis are presented on 105 pages, contain 21 tables. The list of references includes 89 sources.

Research methodology: in our work we used general scientific (synthesis, analysis, deduction, induction) and special (application of exogenous hormones on plant objects, culture of isolated embryos, field experience, seed stratification) biological research methods. The main biological methods were laboratory and field experiments, based on experimental, comparative and descriptive methods. Analysis and statistical processing of the data obtained were carried out.

The main pedagogical methods were: verbal, practical and visual. The verbal method is presented in the form of lectures, the practical method is presented in the form of laboratory and practical works, during which students study some phenomena with the help of certain materials and equipment, the visual method implies the use of illustrations (tables) and demonstrations (conducting experiments) in the learning process.

Structure of the work and explanation of the logic of its construction: acquaintance with literary sources on the problem, proposing hypothesis and predicting research results. The experimental part of the work was carried out, the analysis of the results was carried out, an elective course in biology for 10 - 11 grades was developed.

CHAPTER 1

LITERATURE REVIEW. SEED FORMATION

CONDITIONS AND DORMANCY DEPTH

§ 1.1. Effect of photoperiod and light on seed germination

The phenomenon of photoperiodism is the effect of day length on the rate of development, flowering and fruiting of plants. All plants are divided into short-day plants, which flower more quickly when the day is short (less than 12 hours), and long-day plants, which develop actively when the day is longer than 12 hours. The nature of the photoperiod has a major influence on the properties of the seeds, such as their ability to germinate. The photoperiod also affects the size and colour of the seeds, as well as the thickness of the seed coat. Maturation of seeds on a long 16 to 18 hour day leads to smaller seeds with a thicker rind and therefore a deeper dormancy than on a 6 to 8 hour day. The length of the day during seed ripening leads to dormancy unrelated to their morphological characteristics. In this case, the effect of the photoperiod on seeds depends on the preservation of chlorophyll in the tissues surrounding the embryo during this period. Chlorophyll content and its rate of decline determine the content of phytochrome forms. Seed germination is influenced by the length of the day and the number of periods of light. For some seeds a single light exposure is sufficient, for others light exposure must be repeated over several cycles (Nikolaeva, 1999).

Light sensitivity is a very variable trait and is influenced by various factors: conditions during seed germination, post-harvest seed treatment, growth conditions of the mother plant, hereditary genotype.

Swelling conditions: Seeds that germinate equally well in light and darkness can become light-sensitive if they are swollen at a slightly higher than optimum temperature. For example, lettuce seeds germinate by more than 90% when kept in the dark for 72 hours. At 24o C, but only 32% of the seeds germinated at 29o C. However, with continuous low-intensity white light, germination at 29o C reached 90% again. It was also shown that the normally light-sensitive seeds of the Grand Rapids lettuce variety germinated at 24o C at almost 100 % and at 29o C at 95 %. At the same temperatures in the dark, germination was 65 and 15%, respectively. The results of a comparative study of different seeds showed that super-optimal temperatures can induce a demand for light in initially insensitive seeds and increase this demand in light-sensitive seeds.

Water stress, due to limited moisture supply to the seed during swelling or the high osmotic potential of the solution in which swelling occurs, can induce light sensitivity in the seed and reduce the percentage of germination in the light or in the dark. Soaking in 0.1 M polyethylene glycol in the dark at 20o C was found to reduce germination of Grand Rapids seeds by up to 11%. After a brief exposure to red light, germination increased to 81%. Germination of seeds soaked in water was 86% in the dark and 98% after exposure to red light. The effect of swelling conditions, temperature and osmotic potential of the solution in which the eucalyptus seeds were soaked in light was also investigated. It was found that light increased the germination of seeds incubated for 5 days at 30o C up to 90% compared to 50% in dark controls. In the presence of osmotic (polyethylene glycol), the percentage of seed germination at ^{30o} C was 50% in the light and was close to zero in the dark. B

germination of seeds in the dark at 15 to 30o C was negligible in the presence of osmotic, but when the incubation period was doubled, seed germination at 20o C was slightly higher than at 30o C.

However, under the same conditions in light, the percentage of seed germination increased with increasing temperature (30o C, 70%; 25o C, 50%; 20o C, 30%; 15o C, 10%). In contrast, light accelerated germination of oat and Grand Rapids seeds at high moisture levels but slowed germination at low levels. On the basis of these data W. Vidaver suggested that light can in some cases enhance the inhibitory effect of low water potential in seeds, and in other cases it helps to overcome this inhibition. The stimulating or inhibiting effect of light on seed germination depends on temperature, water potential and seed species.

It was found that light sensitivity can be induced in seeds that germinate easily in the dark by exposing them to far-red light during swelling. In Maya lettuce seeds of Noran and Queen varieties it was shown that exposure of seeds to far-red light continuously for 22 hours induced almost complete quiescence in them at 22o C. If exposure to far-red light was followed by brief exposure to red light, about 100% of the seeds germinated within a day. The same effect was observed when the seeds were incubated at 37°C instead of being exposed to far-red light (this applies to most light-insensitive lettuce varieties). It should be noted that in thermo-grown seeds germination was induced at lower doses of red light than in seeds treated with far-red light. W. Vidaver attributes such responses to the presence of alternately working systems, which probably contain not one but several phytochrome populations (Vidaver, 1982).

According to modern concepts (Ustyugova, 2003; Prusova, 2016; Kabashnikova, 2014) the phytochrome system is involved in most of the seed germination responses to light. Researchers (Vidaver, 1982) have put forward several hypotheses explaining the role of the phytochrome system in the regulation of a great variety of seed responses to light. However, a sufficiently complete explanation of the action of phytochrome at the cellular or molecular level is still lacking.

Red light is known to stimulate and far red light to inhibit germination of lettuce seeds. The degree of stimulation or inhibition depended on the batch or variety of seeds of the same species, pre-treatments and/or swelling conditions. Variations in any of these factors can enhance or inhibit germination and thus affect the seed's response to light. External factors, including light, act on germination by influencing endogenous processes that lead to germination or its suppression. It is assumed that there are two opposing systems in the seeds, one that stimulates germination and the other that induces rest. The system that stimulates germination is activated in the presence of phytochrome far red light, phytohormones: gibberellic acid and ethylene, optimum temperatures and sufficient moisture. The rest-inducing system responds to red light phytochrome, high temperatures, water stress and abscisic acid. While swelling conditions are equivalent in other respects, the germination ability of seeds depends on the influence of these two systems. In some seeds, the germination-inducing system initially predominates and may occur in spite of far red light, slightly increased temperature and/or a slight lack of water. When the effect of the systems inducing germination or rest is more or less the same, germination may be inhibited when one or all of these factors are changed. Thus, the effect of any factor depends on the relative influence of

of each of these systems, so the response of seeds to light can vary from vigorous germination to complete quiescence. Consequently, the different germination ability of the different seed types depends on certain environmental conditions and the genetic programme of the seeds themselves.

Germination is somewhat similar to other plant morphogenetic processes which, like germination, are regulated by hormonal and phytochrome systems. Phytochrome and hormones, which are present in other plant tissues, are always present in the seeds as well. Thus, the germination process is regulated in the same way as other growth processes. This regulation consists in the stimulating effect of gibberellins, cytokinins, ethylene and phytochrome far red light, while abscisic acid and phytochrome red light inhibit germination. Since seeds respond to exposure to any of these factors, it can be assumed that the ratio of their activities determines the seed germination response to light. For example, if the activity of abscisic acid is high enough, treatment of seeds with gibberellic acid, ethylene or red light may not induce germination. On the other hand, the threshold level of activity of any inhibitor can be overcome by endogenous activation of the stimulant or by its exogenous introduction into the seed. An example of this is overcoming the effect of red light phytochrome with exogenous gibberellic acid. That is, germination regulation is carried out through the interaction of endogenous hormones and the phytochrome system. Any changes in their activity are reflected in germination.

It should be noted that there are significant differences between the morphogenetic processes of plants and seed germination: even in rapidly germinating lettuce seeds, a dormant state can be induced, in which they will not germinate, despite the presence of optimal conditions. Seeds of many plant species are usually dormant and remain dormant until a disruptive factor occurs. For example, the quiescence of lettuce seeds depends on the integrity of the endosperm and if it is maintained, germination does not occur. Thus, the ability to germinate in some species is limited by the restraining factor from the endosperm.

Two effective ways of inducing dormancy in swollen lettuce seeds have been established: inhibition of germination by far-red light in sensitive seeds and inhibition of germination by incubating seeds at temperatures above 30o C (for sensitive and non-sensitive seeds). The depth of thermokoyage induced by high temperatures depends on the duration of thermal incubation. It was shown that in Grand Rapids salad seeds, it takes at least 10 days to induce dormancy after a preliminary brief exposure to far-red light. At the temperature optimum for germination (20oC), the stimulating effect of exogenous hormones or red light decreased with increasing incubation time. Germination was inhibited even if red light irradiation was used before the application of high temperatures; the final germination in these seeds was slightly higher than in those which received no light or only far-red light at the beginning of the experiment. With the onset of dormancy, germination occurred only if red light was used in combination with another stimulant: gibberellic acid, ethylene or thiourea. It is of interest that any combination of chemicals was only an additional stimulant. Thus, the points of action of ethylene and thiourea are different; neither gibberellic acid nor thiourea regulate germination by stimulating ethylene biosynthesis or activity. Neither thiourea nor ethylene stimulate germination by regulating gibberellin synthesis or activity; neither of the chemical regulators used acts in a manner similar to the phytochrome system.

Stimulation of seed germination by red light is not associated with increased activity or biosynthesis of ethylene and gibberellins.

The endosperm of lettuce seeds is known to act as a germination inhibitor (the endosperm must be removed or disrupted to induce germination of dormant seeds). However, it is not clear whether the change in seed germination response is the result of processes occurring in the endosperm.It is assumed that resting occurs due to the reduced ability of seeds to overcome the available inhibition as a result of the strengthening of inhibitory systems. Complete resting of seeds is a result of combined effect of all endogenous stimulating systems (far red phytochrome, ethylene, gibberellins, cytokinins, etc.) and combined effect of inhibitors (red phytochrome, abscisic acid, limitations imposed by endosperm, etc.).

On this basis, it can be assumed that the phytochrome system mediates the effect of light on germination by functioning in concert with other morphogenetic regulatory systems existing in the seed.

Seeds are insensitive to light when the combined stimulatory effects of endogenous systems can induce germination regardless of the phytochrome status. Light sensitivity appears when the equilibrium between the stimulating and inhibitory systems is determined by the phytochrome state. Seeds are dormant if phytochrome cannot change the equilibrium in favour of the stimulatory systems.

Seed light sensitivity results from the activity of the phytochrome system. The function of phytochrome in seed germination is similar to its function in other processes of morphogenesis. As in other processes, phytochrome can stimulate or inhibit seed germination, it depends on the seed species and the activity of other growth regulators. In seeds whose germination is stimulated by light, far-red phytochrome induces germination and red light phytochrome suppresses it. In seeds whose germination is inhibited by light, the roles of phytochromes are exactly opposite. The role of phytochrome in light-inhibited seeds seems to be similar to its role in the induction of flowering in short-day plants and light-stimulated seeds with long-day plants. The apparent role of phytochrome in germination reflects its interaction with other endogenous regulatory systems, mainly hormonal, but partly mechanical as well (Vidaver, 1982).

§ 1.2. Effect of chemical treatment

Stimulation of fruit formation (pollination, fertilization or external conditions) is associated with a significant increase in phytohormone levels in the ovaries, mainly auxins and gibberellins, which has been confirmed by experiments on obtaining parthenocarpic fruits by treating the ovaries with gibberellins and auxins. Results have been obtained for many cultivated plants from the Solanaceae and Pumpkin families. Similar effects are caused by gibberellin treatment of fruit trees, e.g. apple trees, and cytokinins treatment of ficus (Nikolaeva, 1999). A positive effect of gibberellin on cherry fruit was found. After treatment of flowering plants, the formation of full-fledged ovary, increased fruit size, accelerated ripening time and improved biochemical quality parameters were observed (Prychko, 2014). It was noted that spraying buckwheat plants with epibrassinolide solution in the phase of budding led to acceleration of fruit ripening (Kolotovkina, 2003). Spraying buckwheat plants in the phase of budding with epibrassinolide solution increased seed germination by 39% and seed weight per plant by 37 - 38% (Mishina, 2016). An increasing number of studies are now being carried out with synthetic and combined growth regulators. It was found that natural pollination with etrel increased the yield of courgette plants by 12 - 15% and field germination by 5 - 6% compared with artificial pollination (Kirillova, 2015). A phytohormonal preparation such as abrolin, when applied to apple ovaries, increased yield by 14% and fruit weight by 24% (Barabash, 2017). The effects of growth regulators such as Vympel and

"Biolan" on grape yields. It has been found that Biolan increases the yield by 22 - 26% and Vympel by 18 - 26%, depending on the grape variety (Shepeleva, 2018). It has also been noted that

"Biolan boosted buckwheat yields by an average of 23% (Mishina, 2016).

As a result of a study of the effects of Zircon preparations, "Epin-extra."

"Siliplant-U", and "Siliplant-D" on fruit-bearing vine plantings are shown that the studied growth regulators contribute to an increase in shoot length (by 21% when "Epin-Extra" is applied, by 51% - "Zircon"), berry weight by 22 - 24%, number of berries in the bunch by 2 - 47%, bunches weight by 2 - 46%. The efficiency varies depending on the concentration of preparations, number and term of treatments, as well as varietal specificity (Borisova, 2018). Treatment of buckwheat plants with "Zircon" increased the weight of grain by 18 - 24%, and the application of the solution "Epin-Extra" increased the buckwheat yield by 84% (Mishina, 2016). The application of the growth regulator 'Poteitin' increased potato yield by 17% compared to untreated, and starch yield by 24% (Ustimenko, 2018). It is noted that the use of the growth regulator "Energy - M" in the cultivation of onions stimulated the growth and development of plants. The biomass increased compared to the control by 26%, the average weight of bulbs by 28%, and the yield increased by 47% (Kalmykova, 2018). The study of the effect of the growth regulator "Energy - M" was also conducted on tomatoes. It was found that stem growth by varieties was 6 - 13% higher than in the control. The time of sprouting was reduced by 5 days to 13 days. The yield of tomato increased by 25% from the control variant. The best performance was obtained when using the growth regulator "Energy-M" throughout the growing season (Petrov, 2018). Presowing soaking of sweet pepper seeds with "Energy-M" solution contributed to an increase in seed germination energy by 10 - 18% and their germination by 5 - 12%. The complex treatment of seeds and plants of sweet pepper significantly increased the yield - by 134 - 150% depending on the variety (Kalmykova, 2017). It is known from literature sources that treatment of flowers of Asiatic Chistotelium and Chinese Buttercup with benzylaminopurine (10 mg/l) and epibrassinolide (0.01 mg/l) led to increase of endogenous cytokinin levels in ovaries and formation of seeds with less dormancy depth. The influence of phytohormones on the dormancy depth of emerging seeds was suggested through a change in the hormonal status of the ovary (Dulin, 2003). Pollen is a rich source of phytohormones in nature. In the developing fruit, hormones are actively produced by the ovary and then by the seeds. Cytokinins contained in the young endosperm are known to stimulate berry growth in grapes. However, all questions about the relationship between fruit development and phytohormone dynamics have not yet been resolved (Nikolaeva, 1999).

§ 1.3. Influence of ecological and geographical conditions on properties seeds

A special area of research is the study of the relationships between the properties of seeds and the environmental conditions under which the seeds are exposed in nature. In this case it is important that the reserve stock of seeds is preserved in the soil and the conditions that promote seed germination coincide with the conditions that ensure further development of seedlings. In nature, seeds are exposed to conditions that are a complex of closely interrelated climatic and soil factors. These factors are determined by the geographical diversity of territories, seasonal climate variations, as well as microclimatic and soil features. For example, daily temperature variations, diversity of substrate properties, degree of seed immersion in the substrate, etc.

The distribution of plants across geographical areas is often determined by the properties of the seeds. For example, species with deep physiological dormancy are not found in the tropics, due to the need for cold stratification to disturb it. The dormancy state and the ability to tolerate adverse conditions in tropical seeds depend on the presence of inhibitors and/or structural features (underdeveloped embryo and/or presence of hard covers). Complete dormancy is characteristic of mangrove forests: the seeds germinate on the mother plant. In deserts and arid areas the phenomenon of hardy seed germination is common. Deserts are also characterised by the presence of a large number of ephemera (Poptsov, 1976; Nikolaeva, 1999).

In the temperate zone, seeds have a wide variety of anatomical and morphological properties and adaptive reactions that contribute to the successful reproduction of plants. The nature and duration of seed germination of different taxonomic groups differ significantly depending on the growing conditions (Nikolaeva, 1999). Seeds of steppe and grassland plants are characterized by a shallow physiological dormancy, and their germination is accelerated (Bespalova, 1980; Borisova, 1994). For northern species with irregular seed regeneration due to climatic conditions, deep seed dormancy is one of the adaptations to harsh conditions. However, many researchers note the high laboratory germination of seeds growing in the tundra (Hodacek, 1974; Nikolaeva, 1978, 1983).

In some genera (e.g. Acer and Salix), seed properties are determined not only by geography but also by the life cycle of the plant. For example, the seeds of early flowering species are dormant and germinate immediately after ripening, while the seeds of many late flowering plants are dormant.

Plant cultivation has a significant impact on the properties of seeds. The seed dormancy of most agricultural plants is less deep than that of their wild relatives. For example, the seeds of the genus Impatiens are characterized by deep dormancy, unlike the ornamental species Impatiens balsamina, whose seed dormancy is shallow and is removed during dry storage. A similar phenomenon can be observed in the cultivated forms of many cereals. The processes of intraspecific differentiation of plants are influenced by seed heterogeneity, which is related either to the population heterogeneity of plants or to the timing of flowering. Seed heterogeneity can be observed even within the same plant on shoots of different order or within an inflorescence (Nikolaeva, 1999).

CHAPTER 2

LITERATURE REVIEW. EFFECT OF PHYTOHORMONES ON GERMINATION OF INTACT SEEDS AND ISOLATED EMBRYOS

The role of phytohormones in seed germination is currently under active discussion (Dulin, 2010; Nefedyeva, 2013; Smashevsky, 2017).

Each of the known phytohormones is believed to be involved in seed germination in some way. The phytohormones are known to be in a bound inactive form in mature dry seeds. Their activity changes during germination. For example, the amount of abscisic acid increases in swollen seeds, the amount of which decreases with the beginning of germ growth, while its inhibiting effect disappears or weakens. This is confirmed by experiments with abscisic acid solution in seeds of different species (lettuce, wheat, different types of mustard). At a concentration of 10-4 to 10-5 M, the acid inhibits germination, while at a concentration of 10-6 to 10-9 M it has no inhibitory effect on seed growth.

Gibberellins are known to stimulate the germination of dormant seeds. There are two points of view on the way of their formation in germinating seeds. According to the first one, the gibberellin synthesis apparatus is established during maturation and begins to function after seed hydration and restoration of cell membrane structure. The second view is that gibberellins in mature dry seeds are released as conjugates during seed germination.

Experiments carried out on seeds of cereals and dicotyledonous plants (peas, beans, sunflower, lettuce, hazelnut) showed that gibberellins appear in concentration - 0.5 mg/l in 2-4 hours after seed treatment and stimulate metabolism and growth of axial parts of the germ. They also enter the hoarding organs (the aleurone layer of the endosperm and/or seed coat) and lead to the formation of hydrolytic enzymes. However, the formation of some hydrolases (phytase and lipase) depends on the presence of the germ but is not controlled by gibberellin.

After the discovery of the effect of gibberellins on enzyme activation, it was thought that this was the role of these hormones in seed germination. However, further studies showed that the formation of α-amylase under the influence of gibberellins occurs after seed germination. Consequently, the involvement of gibberellins in seed germination is related to two directions of action: first, in low concentrations, they stimulate the initiation of germ growth and then, in higher concentrations, they induce the formation of enzymes in the endosperm.

The content of free cytokinins increases significantly before the appearance of the first mitoses. It follows that they are a factor regulating the transition of cells to division.

Ethylene promotes seed germination by stimulating hypocotyl cell stretching. It was found that ethylene stimulates germination of seeds of peanut, lettuce, amaranthus, etc. starting from the concentration of 0.2 - 0.7 µl/l, and has the strongest effect at the concentration of 10 µl/l. In the germinating seeds of beans and cocklebur the release of ethylene at a concentration that stimulates hypocotyl growth was observed.

Treatment with auxin solutions is not known to stimulate seed germination. It was found that a significant amount of indolylacetic acid is contained in the hoarding organs (cotyledons and endosperm) of mature seeds. The highest concentration of auxins is found in the seeds of yellow acacia, rice, wheat and maize. However, with the onset of seed germination, the content of indolylacetic acid rapidly decreases to growth-promoting levels. Meanwhile, there is evidence that when combined with cytokinins, auxins stimulate the initiation of germ growth.

There is also the view that phytohormones are involved in shaping the structural and metabolic readiness of seeds for maturation and are not required for germination initiation. Some researchers believe that only gibberellic acid and abscisic acid control seed germination. This view is supported by experiments with seed mutants *of Arabidopsis thaliana*, tomato and other plants capable or incapable of gibberellin and abscisic acid synthesis, sensitive or insensitive to them. It is believed that gibberellins are essential for seed germination and abscisic acid induces quiescence in developing seeds. However, using these experiments to prove the controlling role of gibberellic acid and abscisic acid in seed germination does not exhaust the problem of phytohormone regulation of seed germination.

As germination proceeds, the seedling develops its own phytohormonal system, allowing the organs of the seedling to become less dependent on the phytohormones of the seed and to regulate growth independently. Cytokinins are thought to be formed in the root meristem, gibberellins in the young leaves and root extension zone, and abscisic acid in the root sheath. There is limited information on the involvement of phytohormones during seedling growth. It is assumed that abscisic acid inhibits cell division in the root meristem, gibberellins stimulate shoot growth, weaker - root growth. Auxins regulate cell proliferation in the root. In low concentrations they stimulate root growth and in high concentrations they inhibit it. Ethylene has also been found to be involved in growth processes, as in a number of plants seedling growth is accompanied by its release. As the place of cytokinin formation is the meristematic cells of root, they are considered to induce cell division of this organ (Nikolaeva, 1999).

§ 2.1. Phytohormones in dormant seeds. Role of physiologically active substances in dormant seeds.

The reason for dormant seeds not germinating was long thought to be the presence of growth-inhibiting substances in the seeds or their pericarp or a lack of growth stimulants. In recent years it has become known that the dormant state is regulated by a balance of stimulating and inhibiting substances. The chemical nature of phytohormones and their points of action are currently under discussion.

According to the most recent ideas, abscisic acid and phenolic substances are inhibitors, while gibberellins and cytokinins are stimulators (Kuznetsov, 2005; Avakyan, 2012; Aksenova, 2013; Smashevsky, 2017). Indolylacetic acid is also sometimes referred to as a stimulant. In contrast to the views presented above, M. G. Nikolaeva considers phytohormones not as stimulants or inhibitors, but as regulators of growth processes. As depending on the concentration, place and conditions of action phytohormones can act as stimulators and as inhibitors of growth and germination.

Data on the dynamics of phytohormones during seed dormancy are scarce. The conclusions about the action of phytohormone solutions on dormant seeds are not always reliable, because the type of seed dormancy, the role of covers, temperature and other conditions are not always taken into account in experiments. In spite of this, the data accumulated during the last decades give the possibility to understand the role of many phytohormones in seed dormancy and germination. Let us consider the information on the content and action of individual phytohormones in seeds in more detail.

§ 2.2. The role of phytohormones in fruit and seed development

At present, quite a lot of data on the dynamics of phytohormones during the growth and formation of seeds and fruits have been collected, based on which the researchers note general patterns. Despite the diversity of the data presented due to the differences in research objects and methodological approaches, the general patterns can be noted. It has been shown in timothy, wheat, grapes, etc., that a considerable amount of gibberellins and auxins comes to ovary with pollen. They are essential for growth stimulation. This is confirmed by experiments on treatment of ovary with solutions of these hormones instead of pollen. This treatment is insufficient for seed development and causes formation of parthenocarpic fruits. Fertilisation is a prerequisite for seed formation.

As seeds begin to form, they themselves become producers of phytohormones necessary not only for their normal development but also for fruit growth. Free auxins, gibberellins and cytokinins are found in the endosperm at the very beginning of seed development. Their concentration reaches a maximum during the transition from the liquid endosperm to the cellular endosperm, then their content in the endosperm decreases. This pattern has been established in the seeds of different species (rice, maize, peas, apple, etc.). In some species, such as apple and pea, a second, less strong increase in auxin and gibberellin content is observed, occurring not in the endosperm but in the germ. Biologically active gibberellins predominate in the early phases of seed development. Auxin activity peaks at a much later stage of seed formation. Seed maturation and dehydration is accompanied by the transition of auxins and gibberellins to a bound state. Cytokinins are found in both free and bound forms in the mature seeds of several species (conifers, peas, lettuce, apple trees).

Studying the composition of coconut milk and liquid corn endosperm, the addition of which to the culture medium contributed to the successful development of seedpods, was important for understanding the role of phytohormones in the initial stages of seed development. These fluids were found to contain auxins, gibberellins, cytokinins and substances of phenolic nature. However, individual fractions of phytohormones isolated in their pure form have low activity, and only the whole complex of hormones of coconut milk and liquid maize endosperm stimulates cell division (Nikolaeva, 1999).

Experiments on cultivation of oil poppy seedpods allowed us to specify the role of some phytohormones at certain stages of embryogenesis (Pontovich, 1978). It was found that during the period preceding embryo differentiation, when growth is due to cell division, kinetin in the presence of auxin is most important. The presence of gibberellins is necessary for the initiation of embryo axis growth. The placenta, which is a source of phytohormones in the endosperm, and a morphogenetic factor consisting of various phenolic substances, are important for successful development of the seedling. Abscisic acid is also present in small amounts in the seeds of various plants from the beginning of their formation (wheat, cotton, pea, bean, pistachio, etc.). As the seed develops, its concentration increases, reaching maximum values in wheat at the beginning or in the middle of maturity, and in peas and beans before maturity. Seed maturation and dehydration is accompanied by a sharp decrease and/or complete disappearance of free abscisic acid (Berestetsky, 1986; Rafieva, 1990).

It is often noted that a high content of abscisic acid coincides with the time of seed growth arrest. But there is evidence that seed growth and synthesis of stored substances can occur against a background of high abscisic acid concentrations, e.g. in barley and rapeseed. It has been suggested that the main role of abscisic acid is to prevent immature seeds from germinating. Fruit growth is not retarded by high concentrations of abscisic acid. Moreover, in soybean it is faster at high abscisic acid concentrations (Nikolaeva, 1993).

High activity of phytohormones plays an important role not only in the normal development of immature seeds, but is also a prerequisite for fruit growth. However, it should be noted that many aspects of phytohormone involvement in seed and fruit formation processes are not yet fully understood (Nikolaeva, 1999).

§ 2.4. The importance of auxins and abscisic acid in germination
  seeds

Auxins, in particular indolylacetic acid, are an important growth factor. However, auxin treatment did not accelerate germination of seeds or isolated embryos. Exogenous indolylacetic acid stimulated growth of wheat coleoptiles but inhibited germination of mustard and maple seeds (Nikolaeva, 1999). Indolylacetic acid at a concentration of 10 mg/l increased the number of dividing cells in barley rootstocks by 7.9% compared with control. However, at an indolylacetic acid concentration of 100 mg/l there were no dividing cells (Dudin, 2003).

A high content of indolylacetic acid was found in the swollen seeds of apple-tree, maple, linden, ash and others in deep physiological rest. In Tatar maple, the content of indolylacetic acid in dormant seeds reached 24.7 µg per 100 seeds. An extract of this hormone indolylacetic acid inhibited germination of seeds of different species and caused abnormal germ growth. 80% of indolylacetic acid is found in the axial organs of the embryo. Based on this, M. G. Nikolaeva (Nikolaeva, 1999) suggested that abnormal germ growth occurs due to high content and uneven distribution of indolylacetic acid in them. This is confirmed by the fact that physiological dwarfism of embryo growth, both natural and caused by treatment of stratified seeds with solutions of indolyl butyric acid, has complete morphological and anatomical similarity.

Under warm stratification, which does not bring the seeds out of physiological dormancy, the concentration of indole butyric acid in the seeds remains high. But under the influence of cold stratification, it decreases and drops sharply before germination. Indolylacetic acid inhibits germination and disturbs the normal growth of isolated embryos in seeds in forced rest, as well as those characterized by deep or intermediate physiological rest. No definite interaction of indolylacetic acid with other hormones has been found.

Many physiologists (Titov, 2009) believe that abscisic acid is responsible for seed resting. There is evidence of its considerable concentration in fruits and seeds of various plants. The amount of abscisic acid is highest in embryos or their axial organs in seeds in deep rest, abscisic acid has a strong inhibitory effect, so it is considered the main factor of seed resting. However, abscisic acid is contained in significant amounts not only in dormant, but also in easily germinating seeds of yellow acacia and silver maple. The emergence of seeds from deep dormancy under the influence of cold stratification is explained by the decrease of abscisic acid content in them. In apple tree seeds, abscisic acid disappeared after one month of cold stratification; in Tatar maple seeds, its concentration was halved after two months of stratification. However, the concentration of abscisic acid decreases with both cold and warm stratification. Seed emergence from dormancy is not always accompanied by a decrease in abscisic acid concentration in the embryo. M. G. Nikolaeva (Nikolaeva, 1999) suggests that the decrease in abscisic acid concentration occurs both due to its binding and transformation into less active substances, and due to diffusion into the environment. The researchers of the Institute of Plant Physiology named after K. A. Timiryazev K. A. Timiryazev Institute of Plant Physiology found that the content of abscisic acid in eyelets and parenchymal tissues of potato tubers increased with the onset of deep dormancy and sharply decreased when the tubers came out of dormancy (Aksenova, 2013). There is literature data that treatment of cereals (wheat, oats, maize) with abscisic acid solutions under water stress (drought or flooding) contributes to normalization of water exchange, reduces the deviation of physiological processes during stressors and increases the efficiency of reparation processes (Bakhtenko, 2003).

It has been found that during the process of seed maturation, significant amounts of indolylacetic acid and abscisic acid accumulate in the seeds during the deposition of stored substances. In mature dry seeds, these growth regulators are in a bound state, but when the seeds swell, they are released and protect the seeds from untimely germination. Seed germination is accompanied by disappearance or reduction of abscisic acid concentration and reduction of indolylacetic acid content to a concentration favourable for growth. Under conditions that inhibit germination, e.g. low temperature, the concentration of indolylacetic acid in the seed remains high. Consequently, it is indolylacetic acid that is responsible for seed quiescence, while abscisic acid has only an additional inhibitory effect on seeds (Nikolaeva, 1999).

§ 2.5. Role of gibberellins, cytokinins and ethylene in seed germination. Ways of enhancing their action.

Gibberellins stimulate the germination of dormant seeds of many species. The nature and conditions of the stimulating action of gibberellins depend on the type of dormancy and the species characteristics of the seeds. Gibberellic acid has no noticeable effect on exogenously dormant seeds, but it eliminates shallow physiological resting. In seeds with this type of dormancy, germination stimulation is observed under the influence of gibberellic acid solutions in a wide range of concentrations. The optimum concentrations vary according to seed type, treatment conditions and gibberellin applied (Nikolaeva, 1999). Soaking barley seeds in gibberellin solutions was found to stimulate an increase in first leaf length. This index at the concentration of gibberellin solutions of 10 and 100 mg/l was increased by 35% and 69% of control respectively (Dudin, 2003). An important aspect of the action of gibberellins is their ability to stimulate germinal pre-development in seeds with morphophysiological type of dormancy. The nature of gibberellins action depends on the species-specific characteristics of seeds, on the type and depth of morphophysiological dormancy (Nikolaeva, 1999). Thus, E. R. Avakyan et al. found that the seeds of early-ripening rice varieties have increased concentrations of gibberellin, which leads to increased activity of hydrolytic enzymes and high content of soluble sugars. This is what the authors attribute to the better and more uniform sprouting of early maturing rice varieties during the germination period (Avakian, 2012). Experiments carried out on dormant potato tubers by the staff of the Institute of Plant Physiology named after K. A. Timiryazev. K. A. Timiryazev Institute of Plant Physiology showed that gibberellins induce tuber dormancy interruption and initiation of bud germination. A low activity of endogenous gibberellins during dormancy and a sharp increase in activity before bud germination were noted. Gibberellins also stimulated active seedling growth in dormant tubers. N. P. Aksenova and co-workers investigated the joint application of gibberellins with cytokinins. For example, when gibberellic acid and benzylaminopurine were applied together to dormant potato tubers, dormancy was terminated faster than when they were applied separately (Aksenova, 2013). A high synergistic effect was also observed when gibberellic acid and cytokinins acted together on soybean and maize plants. Their positive effect on transpiration intensity, water retention capacity of leaf tissues, photosynthetic apparatus structure, intensity of visible photosynthesis and dry weight accumulation in organs of soybean and maize plants was found (Smashevsky, 2017).

The stimulating effect of gibberellins on seed germination and growth of isolated embryos in many species has suggested that seed dormancy is associated with a lack of this hormone. Thus, the factors that bring seeds out of quiescence should lead to an increase in its content and to an increase in its activity. However, numerous observations on seeds of different species have yielded ambiguous results, which may be partly due to methodological differences and species-specific biology.

The stimulating effect of cytokinins is not observed in all seed species, most often in a strictly defined concentration range specific to different species. Exceeding the concentration causes abnormal growth and inhibition of embryo growth. While the gibberellins stimulate the growth of the germinal axis, the cytokinins induce seedling growth. Kinetin relieves germination inhibition in light-sensitive seeds of tobacco, lettuce, etc., but in complete darkness the effect of Kinetin is weakened. It is active in combination with at least a very short light period. Treatment with zeatin, kinetin or benzylaminopurine solutions (50-500 mg/l) stimulates germination of seeds with a shallow dormancy type (peanut, millet, cocklebur). At a concentration of at least 50 mg/l, cytokinins stimulate germination of isolated seed germs with deep physiological type of dormancy (maple, mountain ash, apple, pear). At a concentration of 500 mg/l, cytokinins promote germination of maple seeds, with prior removal of the pericarp. Treatment of apple seeds with kinetin reduces the time of their cold stratification (Nikolaeva, 1999). Treatment of dormant potato tubers with benzylaminopurine caused interruption of dormancy and initiation of tuber germination (Aksenova, 2013). The effect of benzylaminopurine on the anatomical structure of maize plants was studied. The studies showed that benzylaminopurine-treated plants were characterized by a larger stem and vessel diameter; an increase in radial size and width of conducting bundles (Kuznetsova, 2003). Treatment with cytokinin preparations was found to contribute to normalization of plants under conditions of increased concentration of heavy metals and to enhance the growth of axial organs, especially in cucumber seedlings (Bashmakov, 2003). Treatment of wheat seedlings under salt stress conditions with benzylaminopurine reduced the damaging effects of NaCl on plant growth performance (Avalbaev, 2010).

However, cold stratification of seeds under deep resting conditions cannot be completely replaced by cytokinin treatment.

The action of cytokinins can be markedly enhanced by supplementary treatment of seeds with other substances. For example, combined treatment with gibberellin and kinetin stimulates germination of seeds with a normally developed embryo and in deep physiological rest. It is clear that cytokinins play a significant role in the germination of dormant seeds.

One aspect of the action of cytokinins is their ability (unlike gibberellins) to completely inactivate the inhibitory effect of abscisic acid and phenolic substances. However, the nature of their stimulating action remains poorly understood.

Ethylene is considered by some researchers to be the key germination hormone. Treatment with it eliminates the shallow physiological quiescence of seeds of different species. However, there is evidence that the effect of ethylene on seeds with similar shallow dormancy is not the same. It depends on the species specificity, seed coat permeability, seed maturity, gas composition, temperature etc. and causes strong stimulation of germination under some conditions and has no effect under others. The stimulating effect of ethylene can be enhanced by increasing the oxygen content. Such an effect is attributed to the fact that oxygen can stimulate ethylene synthesis. Carbon dioxide can also enhance the action of ethylene. However, there are observations that carbon dioxide and ethylene act independently of each other, and in some cases one gas in the presence of the other enhances seed quiescence. The combined application of kinetin and ethylene (etrel) causes a synergistic effect. Etrel enhances the effect of kinetin on seed germination without the pericarp. Their combination stimulates the germination of seeds of cocklebur, maple, etc. On seeds with deep physiological dormancy, etrel has no stimulating effect, but only slightly accelerates the growth of isolated embryos (Nikolaeva, 1999). Treatment with exogenous ethylene increases the resistance of melon seedlings to UV-B irradiation (Borisova, 2003).

There is evidence that seed germination is accompanied by enhanced ethylene production. In dormant seeds there is also a correlation between the rate of emergence from dormancy and the intensity of ethylene production. The inhibitory effect of abscisic acid can be eliminated by treating seeds with ethylene. It is assumed that the stimulating effect of cytokinins is also carried out by enhancing ethylene formation. Thus, the stimulation of seed germination is largely related to the enhancement of ethylene synthesis (Nikolaeva, 1999). The results obtained by N.P. Korableva on potato tubers are directly opposite to the above mentioned. Treatment of dormant potato tubers with ethylene induced abscisic acid synthesis and prolonged the deep dormancy state (Korableva, 1995; Aksenova, 2013).

Thus, the composition of phytohormones unfavourable for germination is characteristic not only of seeds in physiological rest, but also of seeds in forced rest. It prevents untimely germination of seeds during the maturation period or under unfavourable conditions in mature seeds. The formation of the hormone balance necessary for germination in seeds is determined by the activity of the corresponding enzymes. Their low activity can be explained by the maintenance of high content of indolylacetic acid and abscisic acid in dormant seeds as well as the inability to metabolize gibberellins and cytokinins into forms necessary for germination. Favorable conditions of humidity, temperature and oxygen access are sufficient to activate the enzyme apparatus during germination of non-dormant seeds. But seeds in physiological rest are characterised by reduced enzyme activity. Little is known about the nature of the enzymes involved in phytohormone metabolism. For example, a decrease in indolylacetic acid is usually attributed to the action of oxidase and peroxidase (Nikolaeva, 1999)

§ 2.6. The importance of brassinosteroids in plant growth, stability and productivity

Brassinosteroids represent a new group of hormones discovered in the 1970s. They are capable of stimulating plant growth and development in low concentrations, increasing their productivity and increasing their resistance to stress, which characterises them as bioregulatory and environmentally friendly growth regulators. At present, thanks to these properties, brassinosteroids have found wide practical application in plant growing. The use of bioregulators is necessary for the high efficiency of agriculture and ecological safety of plant protection agents. More than 40 brassinosteroids are known to date. Many of them show growth-stimulating activity at very low concentrations (10-12 M). New physiological and biochemical functions were revealed in the discovered representatives of steroid hormones, which allow to understand their regulatory role in the plant and to develop science-based technologies of brassinosteroids application in agriculture for increasing plant resistance and productivity. In most cases, two synthetic preparations of brassinosteroids are used in research: brassinolide and epibrassinolide.

There are various ways of introducing exogenous brassinosteroids into plants to achieve a growth-stimulating effect: semi-dry treatment, seed inoculation and soaking, spraying of plants in the tubulation or flowering phase. Semi-dry seed treatment with epibrassinolide and brassinolide resulted in seed germination and seedling growth of barley, maize, wheat, spinach and potato. However, brassinolide has no effect on rice stem growth, and epibrassinolide does not remove the inhibitory effect of chlorcholine chloride in perennial wheat plants.

The study of brassinosteroids functions is necessary not only to establish their role in the phytohormonal system of plants, but also for practical applications in crop production. They are known to modify plant auxin metabolism, enhance the synthesis of indolylacetic acid and inhibit the activity of indolylacetic acid oxidase. Brassinolide increases the concentration of indolylacetic acid in pumpkin hypocotyls, and epibrassinolide increases auxin concentration in potatoes, barley and winter wheat. Brassinosteroids increase the amount of endogenous abscisic acid. The effect of brassinosteroids on vegetative growth of plants is explained by the fact that they first affect the content and activity of auxins and cytokinins, and then the level of abscisic acid. The increase of abscisic acid concentration under the influence of brassinosteroids plays an important role in potato tuber germination processes during long-term storage.

Brassinosteroids affect the regulation of photosynthetic processes. Brassinolide and epibrassinolide increase chlorophyll content in plant leaves and cause an increase in photosynthetic activity.

It is known that the effect of brassinosteroids on the enzyme systems of plants depends on their genotype. The response of different plant species to brassinosteroid treatment is associated with changes in physiological and biochemical processes at the level of the protein synthesis system.

Brassinosteroid preparations have been shown to have an anti-stress effect on various plants. Brassinosteroids contribute to plant adaptation to drought, salinity and hypothermia. Cytokinins are known to maintain plant homeostasis, prevent the destruction of plant cells and increase plant resistance to adverse environmental conditions. Brassinolide and epibrassinolide increase the qualitative and quantitative composition of endogenous cytokinins, having a positive effect on the increase of plant stress resistance. Under the influence of epibrassinolide ethylene release increases, which provides plant adaptation to conditions of drought and salinity (Prusakova, 1996). Epibrassinolide treatment of wheat plants was found to have growth-stimulating and protective effect under salt stress conditions (Avalbaev, 2010). The effect of treatment on epibrassinolide in wheat seedlings under conditions of cadmium stress.

Epibrassinolide was found to reduce the damaging effects of cadmium and had adaptive and protective effects on plants (Allagulova, 2015). Brassinosteroids increase the heat tolerance of total protein synthesis in heat shock by inducing the expression of heat shock protein-synthesizing genes.

Under drought conditions, an increase in tissue water content was found under the effect of epibrassinolide. The increase in tissue water content depends on plant genotype and epibrassinolide concentration (Prusakova, 1996).

The effect of epibrassinolide on growth processes, photosynthesis intensity, water regime and productivity of spring wheat varieties differing in resistance to drought was studied under conditions of water deficit. Depending on the phase of plant development, the method of epibrassinolide administration, and different sensitivity to the bioregulator, different growth activity of plants, an increase in the intensity of photosynthesis, a decrease in the number of stomata per unit area, an increase in leaf tissue water content and an increase in plant productivity were observed. The data obtained indicated the participation of epibrassinolide in the formation of wheat adaptation to water deficit and preservation of its productivity (Golantseva, 2006). Brassinosteroids increase plant tolerance to low temperatures. This is observed, for example, in winter and spring wheat, cucumber and rice plants.

Treatment of barley seeds with epibrassinolide contributed to an increase in seed germination and seedling growth in saline conditions, reduced proline accumulation in leaves and prevented the violation of leaf mesophyll cell ultrastructure. Spraying barley plants with epibrassinolide stimulated leaf and stem growth and increased stem strength at first and second internodes. However, an increase in stem strength was not always accompanied an increase stem diamete r. This is
indicates a qualitative change in the stem tissue, resulting in
Increased lodging resistance of the plants. A positive correlation was found between stem strength and calcium ion intake. The protective effect of epibrassinolide on pea seedlings under oxygen stress was shown. After treatment of plants with epibrassinolide, tissues are able to reduce the concentration of primary products of lipid peroxidation. Consequently, epibrassinolide prevents the oxidative breakdown of lipid components of biological membranes, which increases their activity (Prusakova, 1996).

The effect of epibrassinolide on physiological indicators of spring soft wheat under water stress was studied. Epin increased transpiration and increased leaf area per plant by 27% under conditions of acute water deficit. Epibrassinolide had a stimulating effect on photosynthetic intensity. After treatment of wheat plants at the beginning of the tillering phase, plant productivity increased from 24 to 39% depending on moisture conditions. The application of epin has intensified the production process, which suggests that epibrassinolide is a regulator of water metabolism (Vyugina, 2003). There is evidence that brassinosteroids are used as effective and safe means of protection against phytophthora, black root rot and tobacco mosaic virus.

Most often brassinosteroids are used for stimulation of plant growth and development, improvement of conditions for generative development processes, protective action against many kinds of stress, increase of productivity and improvement of agricultural products quality. The treatment of plants by epibrassinolide during the flowering phase increases the number of seeds in the ear by 20 - 30% by increasing the length of the main shoot and the number of spikelets and grains in the ear, which occurs due to the increased outflow of nutrients from the vegetative organs into the ear. Also, depending on the plant genotype, cereal productivity

was increased by higher bushiness and better kernel filling. Maize seed pretreatment with brassinolide resulted in increased productivity due to better earliness of the main and lateral ears and increased grain weight. The barley yield after treatment with epibrassinolide increased due to the fulfilment of grains, the mass of the main ear, and the mass of grains at the expense of lateral shoots. In soybean plants the number and weight of seeds increased by 10-20%; in buckwheat, depending on the variety of plants, the crop increased by 8-25% due to enlargement of seeds and their number.

Treatment of forage cereals and legumes with epibrassinolide resulted in increased chlorophyll, carotenoids and protein content. The treatment of sugar beet seeds stimulated plant growth and development and enhanced the processes of sugar accumulation by 1.5 - 4%. The increase in yield and sugar content of root crops was due to the stimulation of photosynthesis and the outflow of nutrients into the root crops.

The effect of brassinosteroids on productivity, depending on plant species and variety, was manifested in the formation of reproductive organs, their number, completeness and weight. The greatest increase in productivity of cereal crops was observed when plants were treated with brassinosteroids in the flowering phase.

Thus, brassinosteroids are a component of phytohormonal regulation in various plant species. Interacting with other hormones, they increase the content of abscisic acid and have an ambiguous effect on auxins, gibberellins and cytokinins. Brassinosteroids are involved in the processes of growth and development, photosynthesis, protein metabolism, and the supply of ions to the plant. They have a positive effect on plant productivity, resulting in increased yields and improved quality. This class of compounds has a pronounced anti-stress effect on plants.

Brassinosteroids also have fungicidal properties. A characteristic feature of this class is the ability to act at very low concentrations. Based on the above, it can be concluded that brassinosteroids are highly effective, biorational and environmentally friendly regulators of plant growth and productivity (Prusakova, 1996).

§ 2.7. Effect of salicylic acid on seed germination

Salicylic acid is a relatively recent phytohormone (Romanov, 2002).

An overwhelming number of works are devoted to the study of the role of salicylic acid in heat production and in the formation of plant resistance under the influence of biotic and abiotic factors (Bezrukova, 2001; Vasyukova, 1999; Kolupaev, 2009; Shakirova, 2000). The role of the hormone in salt tolerance (Assaf, 2011; Belova, 2018), resistance to extreme high and low temperatures (Kuznetsov, 2005), resistance to soil pathogens - fungi and bacteria (Metlitsky, 1985; Molodchenkova, 2001; Popova, 2017; Mikhailova, 2018) was shown. The role of the new phytohormone salicylic acid in plant growth and development processes is poorly understood. Data on the possibility of salicylic acid regulation of dormancy, emergence from dormancy and seed germination are sparse and obtained for a limited number of species (Vasyukova, 1999; Shakirova, 2000; Bezrukova, 2001; Belousov, 2005; Kolupaev, 2009;

Assaf, 2011)

Э. R. Avakyan et al., studying early maturing rice varieties, found that treatment of panicles of vegetating plants in lactic-wax maturity phase with salicylic acid solution at a concentration of 300 mg/l causes dormant state. Seed germination in the first 10 days after harvesting is reduced by 32 % (Avakyan, 2012). Treatment of wheat seedlings with salicylic acid reduced the effect of salt stress. It was found that

seedbed   preparation of wheat   seeds
salicylic   acid
contributed to reduced water loss and cell membrane permeability under water stress conditions. Salicylic acid reduced the effects of water stress on sunflower seeds. The effect of salicylic acid on the productivity of early- and mid-early potato varieties under different soil and climatic conditions was studied. The highest yield increase

was observed when salicylic acid was sprayed at a concentration of 100 mg/ml (Fedorova, 2014). Treatment of seeds and vegetative plants of flax - dolgan flax by salicylic acid solution contributed to the increase of yield and quality of products, increase of resistance to adverse environmental factors (Belopukhov, 2003). The protective effect of salicylic acid on wheat seedlings under cadmium stress was noted (Maslennikova, 2013). The action of exogenous salicylic acid was found to help in the formation of resistance of pepper plants to UV-induced oxidative stress. Salicylic acid stimulated the activity of antioxidant defense enzymes (Mahdavian, 2008).

CHAPTER 3

EXPERIMENTAL PART

§ 3.1. Objects and methods of research The objects of our work were:

1. Isolated germs and seeds of Manchurian ash (*Fraxinus mandshurica Rupr.* - seed dormancy type - morphophysiological). Time of fruit collection: late October, after leaf fall. Germs were isolated from seeds pre-soaked for a fortnight in refrigerator conditions.

2. Isolated embryos and seeds of felted cherry (*Microcerasus tomentosa (Thunb.)* EreminetJuschev-deep physiological type of dormancy). The time for harvesting Microcerasus tomentosa fruit is in early July, after ripening. Germs of Microcerasus tomentosa were obtained from seeds of two years' dry storage.

3. Isolated germs and seeds of Sargent's wort (*ViburnaceaesargentiiKoehne* - seed dormancy type - morphophysiological). The fruit is harvested in mid-September, after ripening.

4. Isolated germs and seeds of Manchurian apricot (*Armeniacamandshurica (Maxim.)* - intermediate physiological dormancy type). The time for harvesting the fruit is early August, after ripening.

Field and laboratory experiments were conducted in 2016 - 2018 in the laboratory of plant physiology at the Pedagogical Institute of Pacific State University and at a dacha site near Telman Lake. In the field experiment, we used seeds of 8 months dry storage; the embryos isolated from them were soaked in solutions of salicylic acid and gibberellic acid, applying them separately and together. The treated germs were sown on ridges to a depth of 4 cm, seed spacing was 15 cm and row spacing was 30 cm. Field germination was determined after one month.

Isolated embryos were placed for germination in Petri dishes on filter paper moistened with growth regulator solutions: salicylic acid, gibberellic acid, epibrassinolide and indolyl butyric acid. Experimental variants included both separate and combined growth regulator treatments. Distilled water was used as a control. Germs were germinated under daylight illumination at 20-22 ^{o}C, and morphometric parameters were recorded on the 12th day. Flower treatment of mother plants was carried out by spraying with aqueous solutions of growth regulators in the dacha area. In addition to the growth regulators listed above, benzylaminopurine, chlorocholine chloride and etrel were used to treat the flowers.

Germ emergence from dormancy was assessed by the following traits: seedling chlorophyll synthesis, seedling opening, epicotyl, hypocotyl and root growth.

The linear parameters (length of axis, epicotyl, hypocotyl, root) were determined with millimetre paper, the morphometric parameters of the seedlings were measured with a centimetre tape.

Weight determination was carried out using torsion scales (WT, Poland). The data were statistically processed. The tables show the arithmetic mean of several laboratory and field experiments and their standard errors. Repetition of experiments was five times, biological repetition was two times. Analytical repetition was twice. The arithmetic mean (M) and the standard error of the mean (m) were determined. A coefficient of reliability between the two means was calculated; if it was less than 2, the difference was considered unreliable (Klein, 1974).

§ 3.2. Effect of flower treatment of the mother plant on

resting and seed germination

One of the ways to study the role of phytohormones in a particular process is to treat plants or their organs with exogenous hormones. As shown in a number of studies (Puzina, 1999; Sakhabutdinova, 2002; Likhacheva, 2004; Maksimov, 2004; Seldimirova, 2017) treatment with exogenous hormones can lead to increase of their concentration in tissues and change the hormone ratio. There is evidence that hormonal treatment of plants during the flowering phase affects the generative processes and affects the quality of forming seeds (Aliev-Leshchenko, 2015; Mukhina, 2016).

We have attempted to elucidate the effects of hormones alone and in combination on the establishment of resting hormones in the early stages of fetal and seminal formation.

The dormancy state of the formed seeds from Manchurian ash flowers treated with growth regulators was evaluated by the intensity of germination of isolated embryos (Table 1). In the control variant, one fifth of the isolated embryos germinated after two weeks. Seeds had the same deep dormancy in the variant with flower treatment with Etrel. All other growth regulators used: epibrassinolide, indolyl butyric acid, benzylaminopurine, gibberellic acid and chlorcholine chloride contributed to the formation of seeds with a shallower resting depth - the number of germinated isolated embryos varied from 50 to 100%. The highest number of seeds with shallower dormancy was formed in the variants with epibrassinolide and benzylaminopurine. A relationship was found between epibrassinolide concentration and seed dormancy depth. In the variant with a concentration of 0.01 mg/l and 0.1 mg/l, 70% of isolated embryos germinated. In the variant with concentration of 0.5 mg/l all germinated seeds germinated.

Table 1.

Effect of treatment of flowers *of Fraxinus mandshurica Rupr.*

phytohormone on the formation of dormancy depth of isolated germs
(seeds of the 2016 harvest)

Option	Sprouted germs	
	%	% of control
H2O	20	100
EPB 0.01 mg/l	70	350
EPB 0.1 mg/l	70	350
EPB 0.5 mg/l	100	500
BMI 100 mg/l	50	250
BAP 50 mg/l	70	350
GB 1000 mg/l	60	300
CCC 1500 mg/l	60	300
E 150 mg/l	20	100

The effect of Microcerasus tomentosa flower treatment on the dormancy depth of the embryos of the formed seeds is shown in Table 2. In the control variant after two weeks all isolated embryos continued to be in the dormant state. In the variant with treatment of flowers with salicylic acid all embryos came out of the dormant state. Of these, 60% showed chlorophyll synthesis by one cotyledon, 10% by two. Hypocotyl and root growth was observed in 40% and epicotyl growth in 10%. In the variant with treatment of flowers with epibrassinolide, 90% of embryos emerged from the dormant state. Of these, in 80% of them, chlorophyll synthesis was observed in one seedling, in 30% - in two. Hypocotyl growth was observed in 70% of the seeds, and epicotyl and root growth in 50%. In the variant with combined application, 20% of embryos came out of dormancy, in which the synthesis of chlorophyll of one seedling took place, hypocotyl, root and epicotyl growth and chlorophyll synthesis by the two cotyledons were not observed. Storage of Microcerasus tomentosa seeds in a dry state at room temperature for two years resulted in a reduced germinal dormancy. Indeed, while embryos isolated from seeds after two months of dry storage did not germinate within 21 days, those isolated from seeds of two years' storage had a shallower resting depth and most of the isolated embryos germinated after 14 days.

Table 2.

Effect of treatment of the flowers of the mother plant of Microcerasustomentosa cherries on the dormancy depth of the formed seeds (21 days)

Option	Sprouted germs, %	Synthesis chlorophyll by seedlings		Hypocotyl growth, %	Root growth, %	Epicotile growth, %
		one	two			
H2O	0	0	0	0	0	0
SK 10-4 M	100	60	10	40	40	10
EPB10-7 M	90	80	30	70	50	50
SC+EPB 10-7 M	20	20	0	0	0	0

As shown in Table 3, treatment of the flowers of the mother plant Microcerasus tomentosa had an effect on the morphometric parameters of the formed seeds. The application of salicylic acid slightly increased the crude weight of the seeds by increasing the weight of the rind. Epibrassinolide increased seed crude weight by both increasing rind weight and germ weight, but

the combined application of phytohormones did not increase the studied indicators - they remained at the level of the control variant. Germination of isolated embryos on phytohormone solutions affected their emergence from dormancy and growth processes. The application of salicylic acid at a concentration of 10-4 M and epibrassinolide at a concentration of 10-7 M was accompanied by an earlier emergence of embryos from the quiescent state.

Table 3.

Effect of Microcerasustomentosa treatment on the flowers of the mother plant

by phytohormones on some indicators of formed seeds

Option	Raw seed mass		Raw weight of the rind		Raw mass fetus	
	mg	%к	mg	%к	mg	%к
H2O	149,00±0,23	100	112,00±0,45	100	37,00±0,32	100
SC 10-4 M	162,00±0,12	109	124,00±0,40	111	38,00±0,24	103
FTE 10-7 M	174,00±0,25	117	132,00±0,21	118	41,00±0,23	111
SC+EPB 10-7 M	153,00±0,15	103	114,00±0,36	102	39,00±0,45	105

Table 4.

Effect of maternal flower treatment

ViburnumsargentiiKoehnephytohormones on some indicators of formed fruit and seeds.

Option	Fetal weight		Weight of the seed		Seed length		Seed widt h	
	mg	%	mg	%	see	%	see	%
H2O	510,60±0,35	100	39,20±0,20	100	0,72±0,02	100	0,62±0,02	100
SC 10-4 M	607,20±0,75	119	41,80±0,10	107	0,74±0,04	103	0,62±0,02	100
FTE 10-7 M	566,40±0,15	111	43,60±0,30	111	0,75±0,05	104	0,61±0,01	98
SC+EPB 10-7 M	478,60±0,65	94	41,60±0,20	106	0,73±0,03	101	0,63±0,01	102

Treatment of Sargent's wort flowers with the hormones salicylic acid and epibrassinolide promoted an increase in fruit weight, with the application of salicylic acid being more effective with respect to this indicator. However, the combined use of hormones had no effect on the indicator studied.

The positive effect of both hormones was evident in seed weight. The effect of salicylic acid increased by 7% and that of epibrassinolide by 11%. The combined use had an effect comparable to that of salicylic acid.

Linear size of seeds under the influence of hormones did not change, some increase in length lies within the statistical error. Apparently, the increase in seed weight is due to an increase in seed rind, as we noted for seeds of felted cherry tree under the influence of both salicylic acid and epibrassinolide. It is known from literature sources that thickness of seed rind of Chenopodium and Ononissicula species depends on the photoperiod in which the mother plant was grown (Nikolaeva, 1999).

Table 5.

Effect of the flower treatment of the mother plant
ViburnumsargentiiKoehne

phytohormone the ratio of seed length to germ length.

Option	Seed length		Germ length		Ratio of embryo length to seed length
	mm	%	mm	%	
H2O	7,2±0,1	100	1,2±0,1	100	0,17
SK 10-4 M	7,4±0,1	103	1,2±0,1	100	0,16
EPB 10-7 M	7,5±0,1	104	1,2±0,1	100	0,16
SC+EPB 10-7 M	7,3±0,1	101	1,2±0,1	100	0,16

The dormancy depth of seeds with morphophysiological type of dormancy can be characterized by the ratio of underdeveloped embryo and seed length (Nikolaeva, 1999).In seed of Sargent's guelder rose the ratio of embryo length to seed length was 0.17 in the control variant. In variants with treatment of flowers with phytohormones and with their separate and combined application, this indicator did not change, which may indicate the formation of dormancy of the same depth.

§ 3.3. Effect of exogenous hormones on the growth and development of isolated embryos

Dry storage of seeds is known to reduce dormancy depth. This fact was established not only for seeds with a shallow physiological type of dormancy, but also for seeds with deeper types of dormancy (Nikolaeva, 1999). Our preliminary experiments showed that seeds of felted cherry and Manchurian apricot during dry storage of seeds for several months were characterized by less deep dormancy than seeds immediately after harvesting.

According to M. G. Nikolaeva, not only the seed dormancy state but also germ emergence from the dormancy state is under hormonal control. It was found that the concentration of abscisic acid and indolylacetic acid decreases at seed emergence from dormancy. An increase in the concentration of gibberellins and cytokinins was also shown. In addition, seed emergence from dormancy in many plants is accelerated by treatment with exogenous gibberellins and cytokinins (Nikolaeva, 1999; Dudin, 2003; Kuznetsova, 2003; Aksenova, 2013; Smashevsky, 2017). However, the role of hormones such as salicylic acid and brassinosteroids is poorly understood. Data on their effects on plants are sparse and obtained for a limited number of species (Prusakova, 1996; Avalbaev, 2010; Dulin, 2010; Fedorova,

2014; Allagulova, 2015).

Isolated embryos are a good model for studying dormancy and germination. Their advantages are their rapid emergence from dormancy and high germination rates.

Germination of isolated embryos with phytohormone solutions affected their emergence from dormancy and growth processes. The application of salicylic acid at a concentration of 10-4 M and epibrassinolide at a concentration of 10-7 M resulted in earlier germ emergence from dormancy, while the use of indolyl butyric acid at concentrations of 5 mg/l and 10 mg/l, on the contrary, delayed the emergence of seeds from dormancy.

The effect of the different concentrations of salicylic acid on the growth of emerged *FraxinusmandshuricaRupr.* germs is shown in Table 6. All three concentrations used: 10-4, 10-5 and 10-6 M had a stimulating effect on the linear size of the germ axis and the crude weight of the seedling. The optimum concentration of phytohormone was 10-4 M, where the axis size was 53 % bigger than the control variant, and the crude weight of the seedling was 47 % bigger.

Table 6.

Influence of salicylic acid on the growth of isolated germs *FraxinusmandshuricaRupr.* (seed from the 2016 harvest).

Option	Axle length		Raw mass	
	mm	%	mg	%
H2O	10,00±0,01	100	9,50±1,50	100
SK 10-4 M	15,30±0,03	153	14,00±0,80	147
SK 10-5 M	13,60±0,10	136	12,75±0,75	134
SK 10-6 M	13,50±0,15	135	13,00±1,00	137

In a case study on the effect of epibrassinolide on the growth of isolated embryos *of FraxinusmandshuricaRupr.* , all three concentrations had an effect: 10-6, 10-7,10-8 M (Table 7). The variant with epibrassinolide concentration of 10-6 M had the greatest stimulating effect. The combined use of salicylic acid and three concentrations of epibrassinolide resulted in stimulation of axis growth and germ weight. However, the magnitude of the stimulating effect in these variants did not exceed that of the variant with epibrassinolide alone.

Table 7.

Effect of co-application of salicylic acid and epibrassinolide on growth of isolated germs *of Fraxinus mandshurica Rupr.* (seeds of the 2016 crop).

ariant	Axle length		Raw mass	
	mm	%	mg	%
H2O	11,50±0,05	100	24,00±2,60	100
SK 10-4 M	13,00±0,07	113	28,00±2,04	117
EPB 10-6 M	18,00±0,05	138	33,00±1,00	138
EPB 10-7 M	16,00±0,07	123	36,00±1,00	150
EPB 10-8 M	15,00±0,10	115	46,00±2,04	192
SC+EPB 10-6 M	16,00±0,06	123	54,00±0,01	225
SC+EPB 10-7 M	17,00±0,09	131	59,50±2,50	248
SC+EPB 10-8 M	15,00±0,08	115	54,50±2,50	227

The combined application of stimulating concentrations of salicylic acid and inhibiting concentrations of indolyl butyric acid (Table 8) increased the growth processes of isolated embryos *of Fraxinus mandshurica Rupr. compared* to the variant with indolyl butyric acid 5 mg/l, but some stimulation of axis growth compared to control was observed only in the variant with combined application of indolyl butyric acid and salicylic acid at concentration 10-6 M. Seedling crude weight in the variants with combined application of phytohormones was at the same level as in the variants with salicylic acid alone.

Table 8.

Effect *of* co-application of indolyl butyric acid and salicylic acid in different concentrations on growth of isolated germs *of Fraxinus mandshurica Rupr.* (seeds harvested in 2017).

Option	Axle length		Raw mass	
	mm	%	mg	%
H2O	4,60±0,07	100	30,00±0,50	100
BMI 5 mg/l	3,00±0,10	65	26,00±1,75	87
SK 10-4 M	7,60±0,09	165	46,00±1,25	153
SK 10-5 M	5,30±0,20	115	32,00±0,80	107
SK 10-6 M	5,50±0,10	120	40,00±1,50	133
IMC+SC 10-4 M	4,00±0,09	87	34,00±0,20	113
IMC+SC 10-5 M	4,50±0,08	98	40,00±1,00	133
IMC+SC 10-6 M	5,00±0,10	109	36,00±0,25	120

As shown in Table 9, the combined application of stimulating concentrations of epibrassinolide and indolyl butyric acid to some extent removed the inhibitory effect of indolyl butyric acid on axis growth and crude weight index, but growth of isolated germs *of Fraxinus mandshurica Rupr.* was less than in variants with epibrassinolide alone.

Table 9.

Effect *of* co-application of indolyl butyric acid and epibrassinolide in different concentrations on growth of isolated germs *of Fraxinus mandshurica Rupr.* (seeds of the 2017 crop).

Option	Axle length		Raw mass	
	mm	%	mg	%
H2O	6,25±0,57	100	13,00±0,10	100
BMI 5 mg/l	3,77±0,17	60	11,60±0,40	89
EPB 10-7 M	8,75±0,49	140	18,00±0,20	138
EPB 10-8 M	6,60±1,62	106	15,80±0,20	121
IMC+EPB 10-7 M	4,50±0,72	72	14,40±0,10	111
IMC+EPB 10-8 M	4,40±0,80	70	14,00±0,30	108

Incubation of isolated germs *of Fraxinus mandshurica Rupr.* on gibberellic acid solutions showed that at concentrations of 3 mg/l, 6 mg/l, 12 mg/l phytohormone activated germ axis growth by 112-145%, contributed to accumulation of raw weight by 31-78%. The optimum concentration of gibberellic acid was 12 mg/l. The combined application of epibrassinolide and gibberellic acid was accompanied by an increase in the effect of each hormone on both axis growth and crude weight accumulation (Table 10). Thus, in the variant with epibrassinolide the axis length was 140% compared to control, while with gibberellic acid 12 mg/l it was 245%. The combined application of the hormones resulted in a 310% result. We also observed synergism in the combined effect of phytohormones on crude weight.

Table 10.

Effect of combined application of epibrassinolide and gibberellic acid in different concentrations on growth *of* isolated germs *of Fraxinus mandshurica Rupr.* (seeds of the 2017 harvest).

Option	Axle length		Raw mass	
	mm	%	mg	%
H2O	8,40±0,68	100	12,60±0,87	100
EPB 10-7 M	11,80±0,58	140	16,40±0,68	131
GB 12 mg/l	20,60±1,03	245	19,80±1,02	157
FTE+GB 12mg/l	26,00±1,52	310	25,40±1,33	202

The results obtained indicate that the application of growth regulators during flowering *of Fraxinus mandshurica Rupr.* qualitatively changes the formation of seeds by dormancy depth. It can be assumed that phytohormones such as brassins and salicylic acid are involved in the process of germination of *Fraxinus mandshurica Rupr.* embryos from the dormant state. During germination of isolated embryos, axis growth and crude weight accumulation are regulated not only by individual phytohormones but also by their combined effect.

Preliminary experiments showed that salicylic acid at a concentration of 10-4 M and epibrassinolide at a concentration of 10-7 M had a stimulating effect on Microcerasus tomentosa axis length growth. These concentrations were used to investigate the effect of co-application of the hormones. As can be seen from the data (Table 11), application of salicylic acid stimulated growth of the axis by 1.7

times, while the crude weight remained at the control level. Epibrassinolide application significantly stimulated axis growth in the variant with 10-7 M concentration. The 10-6 M concentration of epibrassinolide inhibited axis growth, while the 10-8 M concentration remained at the control level.

control. All three concentrations inhibited the growth of isolated embryo weight. The combined application of stimulating concentrations of salicylic acid and epibrassinolide stimulated an increase in axial length in all cases except the variant with an epibrassinolide concentration of 10-7 M. Crude weight was similar to control in this version, but crude weight growth was inhibited in the 10-6 and 10-8 M concentrations when salicylic acid and epibrassinolide were applied together.

Table 11.

Effect of co-application of salicylic acid and epibrassinolide in different concentrations on the growth of isolated Microcerasus tomentosa germs (seeds of the 2017 crop).

Option	Axle length		Raw mass	
	mm	%	mg	%
H2O	26,50±0,11	100	156,00±1,00	100
SK 10-4 M	44,30±0,67	167	156,50±1,50	100
EPB 10-6 M	22,50±0,25	85	90,00±0,50	58
EPB 10-7 M	39,50±0,45	149	140,50±0,50	90
EPB 10-8 M	27,60±0,23	104	129,00±2,30	83
SK+EPB 10-6 M	33,00±0,30	124	115,50±2,30	74
SK+EPB 10-7 M	27,00±0,23	102	158,00±2,00	101
SK+EPB 10-8 M	40,00±0,50	151	121,30±3,30	78

Table 12.

Effect of different concentrations of salicylic acid on

growth of isolated Microcerasus tomentosa germs

(seeds from the 2016 crop)

Option	Epicotile length		Hypocotyl length		Root length	
	mm	%	mm	%	mm	%
H2O	0,10±0,01	100	0,30±0,06	100	2,50±0,03	100
SC 5×10-5 M	0,10±0,02	100	0,40±0,04	133	2,00±0,05	80
SK 10-4 M	0,20±0,01	200	0,70±0,05	233	2,50±0,08	100
SC 5×10-4 M	0,10±0,01	100	0,20±0,02	67	1,40±0,07	56

Note: Seeds were soaked in water before germ isolation

(control) and salicylic acid solutions for 3 days.

Table 13.

Effect of different gibberellin concentrations on the growth of isolated *Microcerasus tomentosa* germs (seeds from the 2016 harvest)

Option	Epicotile length		Hypocotyl length		Root length	
	mm	%	mm	%	mm	%
H2O	0,10±0,01	100	0,30±0,02	100	2,50±0,12	100
GB 100 mg/l	0,10±0,01	100	2,00±0,10	667	7,30±0,15	292
GB 500 mg/l	1,00±0,05	1000	3,00±0,08	1000	8,20±0,10	328
GB 1000 mg/l	0,10±0,01	100	2,50±0,15	833	7,90±0,11	316

Note: Seeds were soaked before germ isolation
in water (control) and gibberellic acid solutions for 3 days.

The results obtained indicate that the application of growth
regulators during flowering of Microcerasus tomentosa qualitatively
changes the formation of seeds by dormancy depth. It can be
assumed that phytohormones such as brassinosteroids, salicylic acid
and gibberellin are involved in the process of germination of
Microcerasus tomentosa embryos from dormancy. During
germination of isolated embryos, axis growth and crude weight
accumulation are regulated not only by individual phytohormones
but also by their combined effect.

Table 14.

Effect of different concentrations of salicylic acid on seedling growth *Armeniacamandshurica (Maxim.)* (seeds from the 2016 harvest).

Option	Epicotile length		Hypocotyl length		Root length	
	mm	%	mm	%	mm	%
H2O	1,00±0,10	100	2,40±0,12	100	5,00±0,15	100
SC 5×10-5 M	1,20±0,08	120	3,10±0,15	129	8,50±0,13	170
SK 10-4 M	2,50±0,10	250	3,60±0,10	150	9,60±0,11	192
SC 5×10-4 M	2,50±0,07	250	3,30±0,14	138	8,80±0,09	176

Note: Seeds were soaked in water before germ isolation (control) and salicylic acid solutions for 3 days.

Table 15.

Effect of different concentrations of hibberellin on seedling growth *Armeniacamandshurica (Maxim.)* (seeds from the 2016 harvest).

Option	Epicotile length		Hypocotyl length		Root length	
	mm	%	mm	%	mm	%
H2O	1,00±0,05	100	2,40±0,10	100	5,00±0,10	100
GB 100 mg/l	3,80±0,12	380	4,40±0,13	183	13,2±0,12	264
GB 500 mg/l	4,80±0,15	480	4,80±0,11	200	14,9±0,15	298
GB 1000 mg/l	4,10±0,10	410	2,80±0,14	117	8,50±0,12	170

Note: Seeds were soaked in water before germ isolation (control) and gibberellic acid solutions for 3 days.

§ 3. 4. Effect of hormone pre-treatment of isolated embryos on seedling growth and development under field experiment conditions

Table 16.

Effect of combined application of salicylic acid and gibberellin on germination and morphometric indices of *Armeniacamandshurica (Maxim.)* seedlings (seeds harvested in 2016, field experiment 2017).

Option	Field germination		Stem length		Root length		Weight plants	
	%	% к	see	%	see	%	г	%
H2O	25	100	85,0±0,1	100	23,0±0,3	100	29,2±0,4	100
SC 10-4 M	50	200	90,0±0,3	106	27,0±0,2	117	47,6±0,3	163
GB 500 mg/l	70	280	107,0±0,2	126	38,0±0,5	165	53,6±0,3	184
SC + GB 500 mg/l	35	140	85,0±0,1	100	26,0±0,1	113	32,5±0,5	111

As shown in Table 16, the field germination of isolated Manchurian apricot embryos was only 25% in the control. The remaining embryos formed a root, but the hypocotyl did not grow. Thereafter, such embryos tended to rot. Treatment with salicylic acid significantly doubled field germination; an even greater number of embryos (70%) germinated in the gibberellin variant. The combined application of the two phytohormones increased field germination compared with the control, but this figure was lower than in the variants of application of the hormones separately.

Seedlings grown in the variants with gibberellin were characterised by more intensive stem growth; a slight stimulation of stem growth was also observed in the variant with salicylic acid. The combination of salicylic acid and gibberellic acid had no effect on this index.

The hormones studied stimulated root growth. The most intensive root growth was observed in the version with gibberellin. The combined application of the hormones led to results comparable in magnitude with the variant of separate application of salicylic acid.

The integral indicator of growth processes is plant weight (dry or wet). The use of gibberellin in the experiment increased the seedlings' crude weight by 84%; the use of salicylic acid increased it by 63%. The combination of hormones resulted in a result of 11%, more than in the control, but less than when the hormones were used alone.

The field germination of isolated embryos of felted cherry was 40% in the control. Treatment with salicylic acid doubled the field germination; in the variant with gibberellin 60% of germs germinated. The combination of salicylic acid and gibberellic acid had an insignificant effect, close to the control. The most intensive stem growth was observed in the variant using salicylic acid. Gibberellin stimulated stem growth somewhat less. When used together, the observed effect was comparable to that of gibberellin alone. Both salicylic acid and gibberellin, significantly (twice), stimulated root growth. The combination of hormones resulted in a greater result than in the control, but less than when hormones were used alone, 65%. The use of salicylic acid increased plant weight by 54%, slightly less than gibberellin at 43%. The combined use of hormones did not exceed the results with the separate use of hormones.

The increase in biomass under the influence of phytohormones was the result of their effect on the shaping processes and growth pattern of individual organs. The number of lateral shoots, stem diameter, leaf area, thickness and number of lateral roots increased in variants with the use of hormones (Tables 17; 18; 20; 21). In Manchurian apricot, gibberellin application had the greatest effect on the studied indicators, and to a lesser extent - salicylic acid, the combined application of hormones had an insignificant effect, close to the control. In the case of the Felt cherry tree, on the contrary, salicylic acid had a more pronounced positive effect on the number of shoots, roots and leaf area.

It can be noted that gibberellic acid at a concentration of 500 mg/l had the most positive effect on the considered indicators in Manchurian apricot plants, and salicylic acid at a concentration of 10-4 M had the most positive effect on cherry felt plants.

Table 17.

Effect of combined application of salicylic acid and gibberellin on the morphometric indices of the aboveground part of *Armeniacamandshurica (Maxim.)* seedlings (2016 seed crop, 2017 field experiment).

Option	Number of lateral shoots		Stem diameter		Leaf area	
	piece	%	see	%	cm2	%
H2O	5,0±0,01	100	0,50±0,03	100	14,3±0,15	100
SC 10-4 M	9,0±0,01	180	0,60±0,02	120	26,8±0,44	187
GB 500 mg/l	12,0±0,01	240	0,70±0,01	140	28,6±0,30	200
SC + GB 500 mg/l	6,0±0,01	120	0,55±0,02	110	14,5±0,25	101

Table 18.

Effect of co-application of salicylic acid and gibberellin on root morphometrics of *Armeniacamandshurica (Maxim.)* seedlings (seed of 2016 crop, field experiment 2017).

Option	Number of lateral roots		Root thickness	
	piece	%	see	%
H2O	18,00±0,01	100	0,80±0,04	100
SC 10-4 M	23,00±0,01	128	0,90±0,05	113
GB 500 mg/l	43,00±0,01	239	1,00±0,05	125
SC + GB 500 mg/l	20,00±0,01	111	0,80±0,02	100

The positive effect of gibberellic acid on stem growth of herbaceous and woody plants has been shown in numerous works (Perelovich, 2002; Timofeev, 2009; Ruzmetov, 2017; Yuldashev, 2018). For salicylic acid, data on its effect on plant growth processes are scarce. It is known that the effect of salicylic acid on stem growth depends on species specificity and the concentration used in the experiment. Salicylic acid has been shown to stimulate the growth processes of herbaceous plant seedlings of different species (Dulin, 2010). No data on the effect of salicylic acid on the growth of woody plants were found in the literature. The combined use of hormones can be accompanied by the summation of the effect or, on the contrary, by the reduction of the studied indicators in comparison with the action of individual hormones (Moskaleva, 1985; Dustmamatov, 2005; Kropotkina, 2009;

Sidelnikov, 2013).

Table 19.

Effect of combined application of salicylic acid and gibberellin on field germination and morphometric indices of *Microcerasus tomentosa* seedlings (seeds of the 2016 harvest, 2017 field experiment).

Option	Field germination		Stem length		Root length		Weight plants	
	%	% к	see	%	see	%	г	%
H2O	40	100	8,5±0,1	100	8,4±0,2	100	12,0±0,2	100
SC 10-4 M	80	200	12,5±0,2	147	17,0±0,1	202	18,5±0,1	154
GB 500 mg/l	60	150	11,0±0,1	129	16,5±0,1	196	17,1±0,2	143
SC + GB 500 mg/l	45	113	10,6±0,2	125	13,9±0,3	165	14,3±0,3	119

Table 20.

Effect of co-application of salicylic acid and gibberellin on the morphometric indices of the aboveground part of *Microcerasus tomentosa* seedlings (seeds of the 2016 harvest, field experiment 2017).

Option	Number of lateral shoots		Stem diameter		Leaf area	
	piece	%	see	%	cm2	%
H2O	1,00±0,01	100	0,24±0,02	100	6,70±0,35	100
SC 10-4 M	3,00±0,01	300	0,30±0,05	125	9,80±0,20	146
GB 500 mg/l	3,00±0,01	300	0,30±0,03	125	8,60±0,30	128
SC + GB 500 mg/l	2,00±0,01	200	0,30±0,05	125	7,80±0,40	116

Table 21.

Effect of co-application of salicylic acid and gibberellin on root morphometrics of *Microcerasus tomentosa* seedlings (seeds harvested 2016, field experiment 2017).

Option	Number of lateral roots		Root thickness	
	piece	%	see	%
H2O	19,0±0,01	100	0,3±0,04	100
SC 10-4 M	52,0±0,01	274	0,4±0,05	133
GB 500 mg/l	48,0±0,01	253	0,3±0,05	100

SC + GB 500 mg/l	23,0±0,01	121	0,3±0,02	100

CHAPTER 4

Thematic planning for an elective course

in biology.

"Plant growth and development" (elective course for grades 10 - 11).

Any elective course is not only designed to develop cognitive interests, but also to help you to learn universal learning activities (ULA). The development of the SKills has become one of the most important tasks of the educational process. As far back as K. K. D. Ushinsky wrote: "Children should, if possible, learn independently, and the teacher should guide this independent process and provide the material for it. If 30 years ago the aim of education was to equip students with a certain amount of knowledge, in our rapidly developing world it is impossible to predict what the content of education will be in 10 - 15 years.

It is therefore our task to help the pupil to master universal learning activities. What are they?

The definition of the IKEA is based on L.S. Vygotsky's theory of the system-activity approach to learning. The essence of this theory was to organise such a learning process, in which the main place is given to active and versatile, to a maximum extent independent cognitive activity of a schoolchild. The main points of the activity-based approach are the gradual shift from informational reproductive knowledge to action knowledge.

In other words, universal learning activities (ULA) are the ability to learn, that is, the ability of a person to improve himself or herself and to acquire new knowledge that he or she needs in life. They will enable the growing pupil to navigate easily through the fast-changing realities of modernity.

The skills include goal-setting, planning, correcting the plan, setting a learning objective, working with information, structuring knowledge, analysing and synthesising new knowledge, making cause-effect connections, proving one's judgement. Moreover, learning becomes personal - the pupil becomes aware of the need to acquire new knowledge.

Almost all of these skills should be used when students take the elective course "Plant Growth and Development".

The selection of content and the development of a specific set of the most effective, vivid and interesting learning tasks, especially in extracurricular activities, also plays an important role in the formation of the SKD. All this is possible and feasible within the framework of the developed elective course.

Explanatory note.
The elective course "Plant Growth and Development" examines the mechanisms of physiological processes that take place in the plant organism. This course deals with the current understanding of plant development in order to provide students with an understanding of how the plant organism functions as a holistic system.
Plant physiology is an experimental science, giving students ample opportunity to learn how to answer their own questions about plant growth and development. The programme includes the possibility of laboratory sessions. While they are informative, they do not require sophisticated equipment and can be useful for students' research work.
RelevanceThe development of the elective course is related to the need to provide high school students with the opportunity to broaden and deepen their
knowledge of biology, to gain skills in experimental

activities and experience with biological samples,

assess their readiness to learn by

of the speciality (biology) through experience

The study of specialised disciplines within the framework of

of the chosen direction (FZ, 2012)

The aim of the course is to develop knowledge about the functions

of the parts of the plant organism. To develop the ability to

investigate the studied functions in plant organisms.

Objectives:

1)   To develop an interest in the processes   The life processes

of a plant organism.

2)   Tell students about modern developments in

3)   The study of plant growth and development processes.

Basic knowledge and skills requirements.

Pupils should know:

- The essence of the life processes of a plant organism,

- the main ways of bringing seeds out of dormancy,

Pupils should be able to:

- carry out observations and experiments using seeds,

- Present the results of observations in tables and draw conclusions

from the results

- Measure morphometric indicators and be able to compare the data

obtained

Methodological and technical support for the course:

- tables, drawings, diagrams, photos

- computer support for the programme

- Materials and equipment for laboratory sessions

Materials   и   equipment: Seeds,

   isolated   Seeds, isolated embryos, preparaval needles   ,

   slides   ,   blades,   dishes, Petri dishes, filter paper,

distilled water, syringes, millimetre test tubes.   Petri dishes,

filter paper, distilled water, syringes, millimeter scale

paper,   tweezers,   solutions of salicylic and hibberellic
acids   and epibrassinolide solutions.

Approximate calendar   planning   elective   (16 hours).

Table 22. Combined effects of phytohormones on seed growth and development of far eastern woody plants.

№	Topic of the session	Type of session	Quantity hours
1.	The concept of seed dormancy.	Lecture	1
2.	Determining the type of rest in of seeds of DV plants.	Practical session	1
3.	Ways of removingrest seeds	Lecture	1
4.	Ways dormant seeds of some species of far-eastern trees (ash, snowball, guelder-rose, cherry).	Practical exercise	1
5.	Phytohormones: The concept, functions, features.	Lecture	1
6.	Hormone inhibitors stimulant hormones.	Lecture	1
7.	Using phytohormones in agriculture.	Lecture	1
8.	Receiving the seed isolated embryos	Practical exercise	1

9.	Preparation of solutions.	Practical session	1
10.	Causes of seed dormancy	Lecture	1
11.	Types of trees plants in Khabarovsk.	Excursion	1
12.	Seed structure isolated plant embryos	Lecture	1
13.	Measurement of morpho-metric indicators seeds isolated embryos.	Lecture	1
14.	Statistical processing morphometric indicators.	Practical session	1
15.	Summary of results discussion of the results . Report of the work done.	Seminar session	1
16.	Protection research projects.	Seminar session	1

Total:

Lectures - 8

Workshops

- 5

Seminars -

2 Field trips

- 1

Programme content
The concept of seed dormancy (2 hours).

Types of seed dormancy and the factors responsible for it. Involuntary and organic dormancy. The cause of forced dormancy. Factors causing organic dormancy. Types of organic dormancy. Exogenous rest. Physical exogenous rest. Mechanical exogenous rest. Chemical exogenous rest. Endogenous rest. Morphological endogenous rest. Physiological endogenous rest. Types of physiological rest: shallow, deep and intermediate.

Laboratory work No. 1. Determining the type of dormancy in Manchurian ash.

Methods of seed dormancy (2 hours).

Structural, physical and chemical influences on seeds. Techniques to stimulate germination: scarification, impaction, local damage to seed coatings, shell dissection, germ alienation. Physical factors of seed dormancy disturbance. Temperature. Water. Composition of gas environment. Light.

Laboratory work No. 2. Methods of dormancy of seeds of some species of far-eastern trees (Manchurian ash, Sargent's guelder rose, felted cherry).

Phytohormones: concept, functions, characteristics (3 hours).

The concept of phytohormones. History of the discovery of phytohormones. Phytohormones as factors regulating growth and development of whole plant. Auxins, gibberellins, cytokinins, abscisins, ethylene. Common features of phytohormones. Hormone inhibitors and hormone activators. Influence of phytohormones on cell division and stretching, formation of roots on shoots (cuttings), differentiation of tissues, transition to flowering, dormancy and emergence from dormancy. Use of phytohormones in agriculture.

Obtaining seeds and isolated germs (2 hours).

Laboratory work No. 3. Obtaining seeds and isolated embryos.

Laboratory work work № 4. Preparation of working Preparation of working solutions of epibrassinolide, salicylic acid and gibberellic acid.

Causes of seed dormancy (1 hour).

Exogenous: mechanical obstruction and/or watertight covers, pericarp inhibitors such as white acacia, beetroot, iris and apricot.

Endogenous: germination underdevelopment (GAD), physiological mechanism of inhibition (PMT) of seed germination of different strengths (weak, medium, strong) using apple, ash, ginseng, lily of the valley as examples.

Combined: combination of causes causing exogenous and endogenous rest using linden, plum as an example.

Excursion to Dynamo Park: "Types of dormant woody plants of Khabarovsk". (1 hour).

Structure of seeds and isolated plant embryos (1 hour).

Peculiarities of seed structure: seed rind, endosperm, perisperm, embryo. Relation between size of seed and size of embryo. Structure of the embryo: germinal root , hypocotyl, germinal Kidney, cotyledon, coleoptile, epicotyl. Antacian ring.

Measurement morphometric seeds и isolated embryos (2 hours).

Axial length, dry weight and wet weight of the seed. Epicotyl, hypocotyl and root length. Length, width and degree of colouring of the cotyledons. Presence of anthacyanic ring. Number of rotted germs.

Laboratory work No. 5. Measurement and statistical processing of morphometric indicators.

Option	Green cotyledons		Presence of an anthacian ring	Hypocotyl growth	Root growth	Epicotile growth	Rotten
	1	2					
H2O							
SC 10-^{4} M							

FTE 10-7 M						
GB 12 mg/l						

Option	Axle length		Raw mass	
	mm	%	mg	%
H2O				
SK 10-4 M				
EPB 10-7 M				
GB 12 mg/l				

Debriefing and discussion of the results. Progress report (1 hour).

Effect of phytohormones on the growth and development of embryonic parts. Establishment of common patterns, identification of differences. Dynamics of changes in morphometric indicators. Summary of statistical material.

Presentation of research projects

(1 hour). Topics of research projects:

1. The history of phytohormone discovery and research.
2. Identification of the type of dormancy in the monocotyledonous and dicotyledonous classes. Comparison.

    Example: Identifying and comparing the type of dormancy of Japanese iris (monocotyledonous) and Manchurian ash (dicotyledonous).

3. Advantages and disadvantages of seed dormancy methods.

4. Hormone stimulants and hormone inhibitors.

5. Ecological aspects of seed germination.

CONCLUSION

A comprehensive study on the effects of phytohormones (salicylic acid, epibrassinolide, gibberellic acid, indolylacetic acid) on:

- establishment of dormant forming seeds when the flowers of the mother plant (Manchurian ash, felted cherry, Sargent's guelder rose) are treated;

- on germination and growth rate of isolated embryos when hormones are applied separately and together;

- on the growth and development of seedlings under field experiment conditions.

The treatment of mother plant flowers with phytohormones and growth regulators (chlorcholine chloride) was found to promote the formation of seeds differing in dormancy depth.

Epibrasinolide, indolylacetic acid, benzylaminopurine, gibberellic acid and chlorcholine chloride promoted the formation of less deep dormancy in Manchurian ash seeds. Etrel had no effect on dormancy depth.

Treatment of cherry blossoms with felted salicylic acid and epibrassinolide promoted the formation of dormancy to a lesser depth; the combined application of phytohormones was less effective than applying each individually.

Spraying of the flowers of Sargent's wort with salicylic acid, epibrassinolide and their mixture had no effect on the dormancy depth of the formed seeds.

Another aspect of our work was to study the effect of phytohormones, separately and in combination, on the rate of seed emergence from dormancy in a model system of isolated embryos. We studied the effect of epibrassinolide, salicylic, gibberellic and indolylacetic acids on the axial length and crude weight of isolated embryos of Manchurian ash and felted cherry. Epibrassinolide, salicylic acid and gibberellic acid were found to stimulate and indolylacetic acid to inhibit the growth of axial length and crude weight. The combined application of salicylic acid and epibrassinolide on isolated embryos of Manchurian ash and felted cherry resulted in an effect comparable to that of one of the hormones.

The application of indolylacetic acid on Manchurian ash embryos inhibited emergence from dormancy and inhibited growth processes.

The use of salicylic acid and epibrassinolide against a background of indolylacetic acid partially reversed the inhibitory effect of auxin.

The combined application of gibberellic acid and epibrassinolide resulted in a synergistic effect of the phytohormones on the growth processes of isolated Manchurian ash germs.

The data obtained may indicate that not only gibberellins and auxins, but also epibrassinolide and salicylic acid are involved in germ emergence and initial growth processes.

Under field experiment conditions on Manchurian apricot and felted cherry plants the data on the effect of salicylic and gibberellic acids on the isolated embryos germination processes were confirmed and the results on the seedlings growth processes were obtained. The effects of salicylic acid and gibberellic acid and their combination increased field germination, stem height and diameter, root length and plant weight.

The combined use of hormones did not exceed the effect of one of the hormones. This may indicate that the positive effect of individual hormones on growth processes is related to a change in the concentration of the endogenous form of the other hormone. The effect of each hormone takes place within strictly defined concentration limits: low concentrations, as well as high concentrations outside of these limits, not only do not stimulate, but inhibit the process. If one of the applied hormones contributes to an increase or decrease in the concentration of another hormone, a situation may arise in which the joint application of hormones does not lead to a positive effect. An elective course in biology for grades 10 - 11 on "Plant Growth and Development" has been developed, allowing students to participate in the development of dormancy and seed germination problems of far eastern plants. The guiding documents implementing the Federal Law on Education were used. In conclusion, it should be noted that the topic we considered has prospects for further development. It is necessary to continue studying the role of hormones in formation of morphophysiological type of dormancy in Chinese lemongrass and Eleutherococcus pricklypodium plants; start research of growth regulators influence on formation of physical dormancy in such Far East plants as Ussuri soybean, Amur maakia, yellowing Sophora. For the elective course, we will continue to develop methodological guidelines for students involved in scientific and research activities on the basis of this course.